Integer Operations
As envisioned by AJ Thomas

Blatant Mathematics

Integer Operations
Copyright © 2017   AJ Thomas

Self-Published by AJ Thomas: Blatant Mathematics
Philadelphia, PA

Illustrated by Katie Joy Nellis
Printed by CreateSpace in the United States of America

ISBN-10: 0-9985134-0-7
ISBN-13: 978-0-9985134-0-9

All rights reserved. No part of this publication may be reproduced, transmitted, or stored in any form without prior permission from the publisher.

Introduction

In a shadowy forest hides
the dashing Princess of Thieves,
wearing garments spun so the eyes
lose their prey in the leaves,

surrounding herself
with allies on a mission;
they call themselves,
"Masters of Rare Acquisition."

Far to the East,
a young Queen reigns,
donning flowery orange
from shoulder to train.

She sends forth knights
to blot out the stain
of all who pursue
some dishonest gain.

Part 1: Crossroad Encounters

And so, in this land,
when strangers meet
at the cross of roads,
intersection of streets,

the colors they sport
dictate how they greet,
whether smiles and hugs
or weapons unsheathed.

If a band of vagabonds,
blazing in green,
joins with some fellows
hungry and lean,

they all raise cups, shout,
"Down with the Queen!"
and count themselves blessed
to be part of a team.

In a likewise manner,
when a group of few
orange clad soldiers
joins with a crew

of fellows-at-arms,
no harm will befall,
for every man present
upholds the Queen's Law.

It's a dangerous day
when opposites meet,
weapons drawn,
a brawl in the street.

If thieves overwhelm
the Monarch's elite,
each soldier that falls
takes with him one thief.

A steep price to pay,
not time much to grieve,
count those still alive,
split treasures, and leave.

Should lawmen outnumber
criminals at a crossing,
arrest every man, woman,
even the offspring;

each bandit escorted,
secured against flight,
to a round tower prison
by one noble knight.

So the company of soldiers
patrolling that sector
ebbs by the count
of thieves they've sequestered.

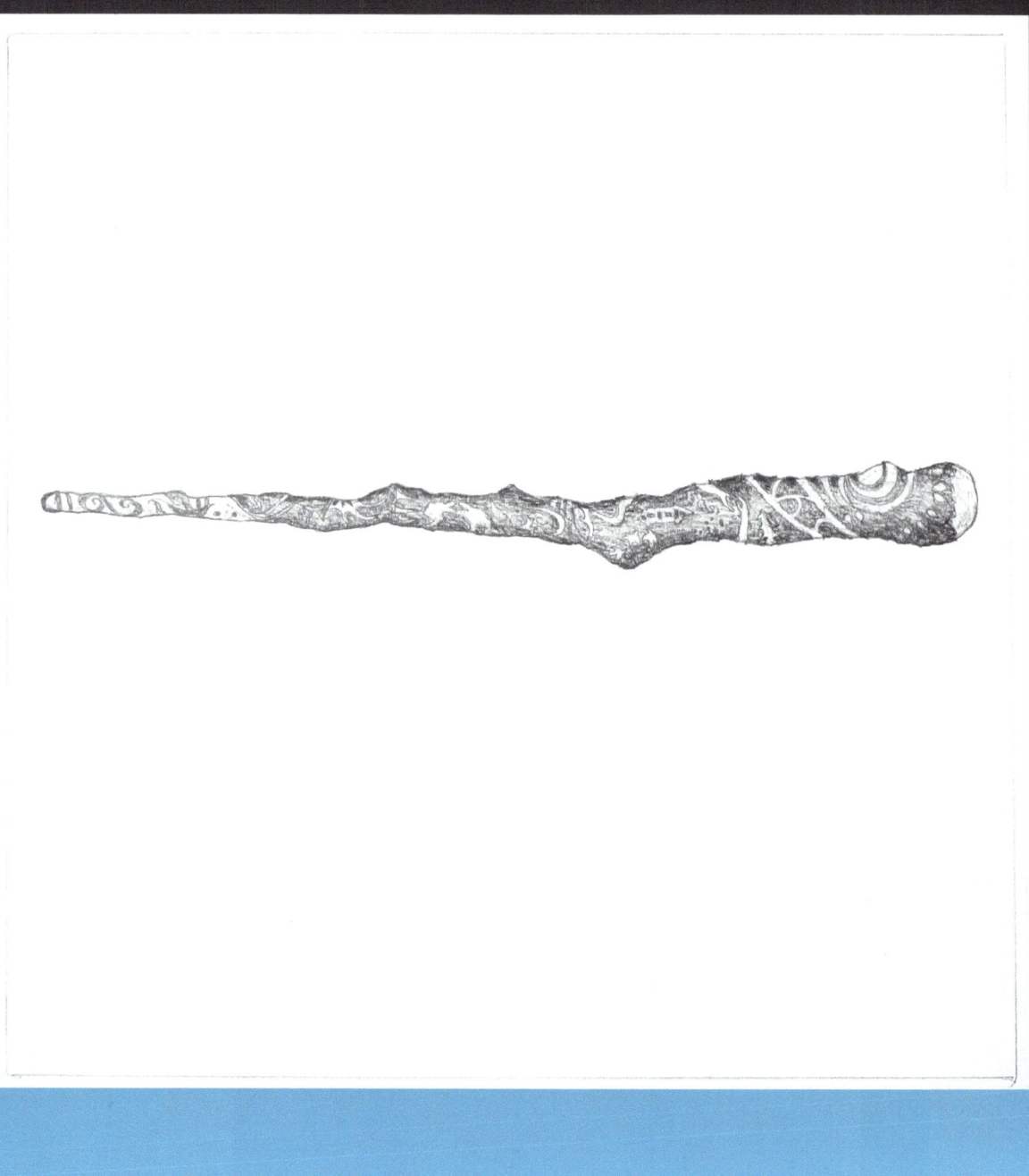

Part 2: A Wavering Wand

There's a madman loose
in the mountains nearby,
with a magical wand
and a patch on one eye.

He delights in mischief,
chaos, his game,
changing virtue to vice
with a flick of the cane,

or vice to versa,
to drive us insane.
We can't trust our eyes
when he switches the frame.

If the Mage hangs about
when opposites meet,
intent on bloodshed,
weapons unsheathed,

then he targets the right
with a devious spell
that changes their robes
and allegiance as well.

What would've been slaughter,
to each group's surprise,
instead is occasion
to celebrate lives.

Dare not think it noble,
this bringing of peace,
and so misconstrue
the Devious Beast.

For if friends approach
when this man hangs around,
he casts the same spell
on the second group found.

Instead of warm greetings,
now enemies fight,
as the grizzled old Mage
laughs with delight!

Beware of the wand,
if you stand to the right,
for the day that you've planned
turns quickly to night.

Part 3: Potions & Poisons

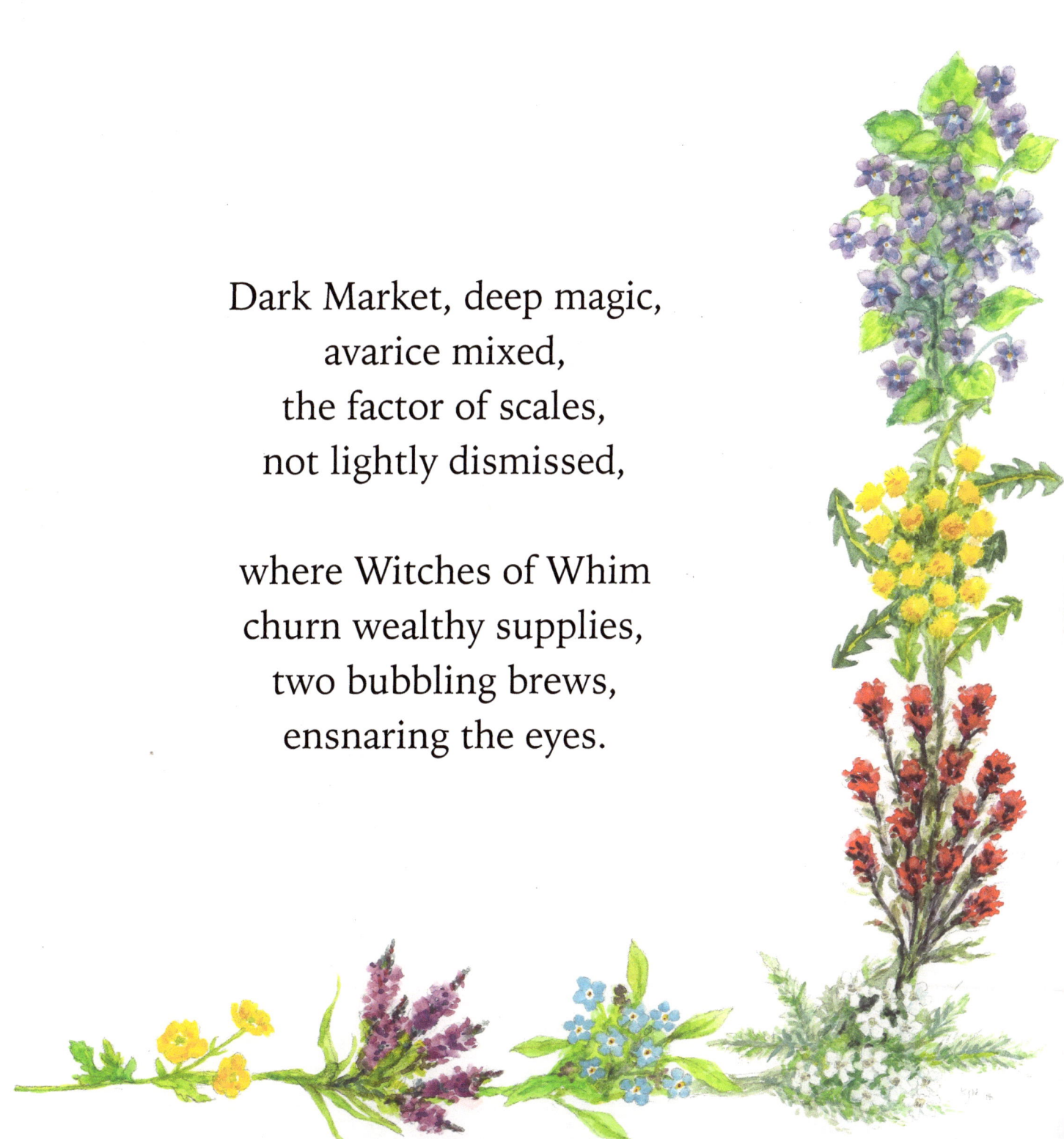

Dark Market, deep magic,
avarice mixed,
the factor of scales,
not lightly dismissed,

where Witches of Whim
churn wealthy supplies,
two bubbling brews,
ensnaring the eyes.

A flask marked with "X,"
of course, multiplies.
Sprinkle its contents
to increase allies.

Where there were just a few,
now stand quite many.
With potent "X5,"
four become twenty!

Or if, through exchange,
one procures the latter,
a draft that divides
enemies when it shatters,

with a dose of "D4,"
twelve guards become three;
and in the confusion,
more bandits can flee.

Part 4: The Mixing of Magics

Some fateful days,
it may come to pass,
the Wizard stands close to
a Witch's brew cast.

The spells become mangled:
"X3" interrupted
by wand-wielding hand,
the outcome corrupted.

Instead of more allies,
the cane and the vase
expand opposition;
"X-3" betrays.

When potion or poison
is spilled on a crew,
to grow or diminish
by Whim Witches' brew

and Wizard waves wand
as he is prone to do,
directed with malice
upon the same few,

the liquid works wonders,
duplicates or divides,
while the Wild One laughs,
reversing the sides.

Conclusion

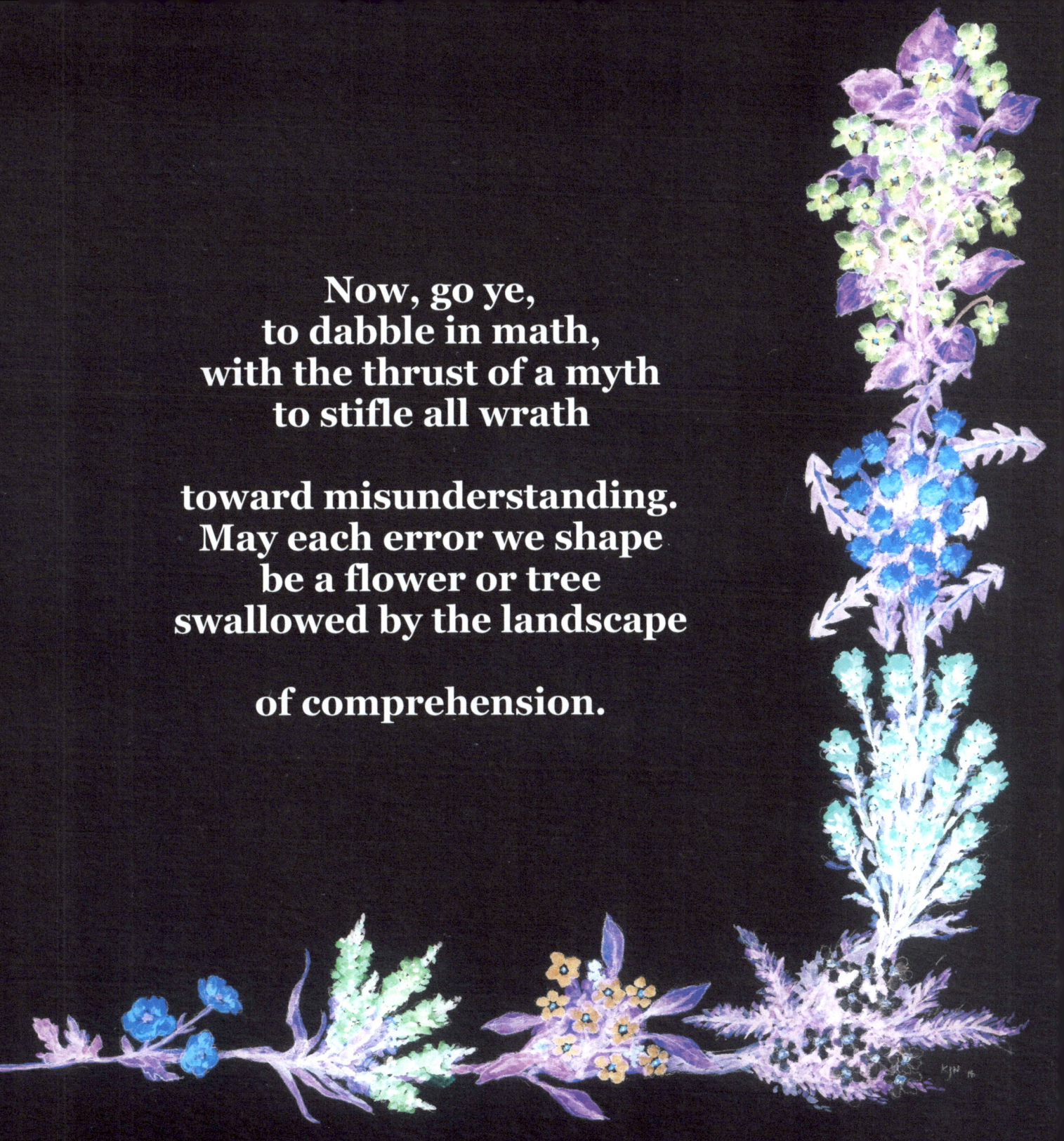

Now, go ye,
to dabble in math,
with the thrust of a myth
to stifle all wrath

toward misunderstanding.
May each error we shape
be a flower or tree
swallowed by the landscape

of comprehension.

Thank you.

To parents, family, friends,
teachers, mentors, colleagues, and coaches,
who've invested in the elevation of this craft,
linguistic and mathematical,
enabling me to publish such a project;

and to the God who gifts
talent, time, and treasure.
May we bear them all
with increasing wisdom.

Katie, thanks for the illustrations;
never could I ever match them!

A. J. Thomas, BA Mathematics and MS Math Learning & Teaching, from Eastern University and Drexel University, respectively.

He has been teaching middle and high school mathematics in the School District of Philadelphia for the past decade.

There may be a follow up "guide" to this mathematical parable, but for now, dance with the pages, pictures, symbols, words, and phrases.

Katie Joy Nellis paints from her home in Coatesville, PA. She studied Art at Gordon College in Massachusetts and in Orvieto, Italy. Since then, she has been working as a freelance artist in a variety of mediums. Although illustration is a recent venture, she credits much of her interest in art to the beautiful pages in books she pored over as a child.

Oil on wood is her favorite medium, "liminal" is one of her favorite words, and she is fascinated by the human face.

Website: KatieJoyNellisArt.com
Facebook: Katie Joy Nellis Art

www.ingramcontent.com/pod-product-compliance
Lightning Source LLC
Chambersburg PA
CBHW060758090426
42736CB00002B/69